Report no. VNTSC-ATMS-05-05

Analysis of Uncertainty in ETMS Aggregate Demand Predictions

by

Scott B. Smith
Eugene P. Gilbo

November 29, 2005

Volpe National Transportation Systems Center
U.S. Department of Transportation
Kendall Square
Cambridge, MA 02142

Executive Summary

ETMS currently makes deterministic predictions of airport and sector demand, based on the conceptual model:

(1) $A(t) = F(t,n) + \varepsilon$ where

$A(t)$ = Actual number of flights for the 15-minute interval starting at time t (interval t)

$F(t,n)$ = Number of flights predicted at time n for interval t (The look-ahead time is t minus n.)

ε = A random error term (currently ignored by ETMS)

Predictions based on this model can be improved by including in the calculation a factor for uncertainty. For example, if flights are frequently and unexpectedly delayed or early, the model can include a deterministic prediction for adjacent intervals (t-15, t+15) along with the current prediction for interval t to improve the prediction of what will actually happen at interval t. The conceptual model that takes into account the immediately adjacent intervals is:

(2) $A(t) = aF(t-15,n) + bF(t,n) + cF(t+15,n) + k + \varepsilon$
(a, b, c, and k are coefficients to be determined)

Regression analysis was used to develop several new models for predicting the number of flights in a 15-minute interval. Models for predicting airport arrivals and the peak number of flights in a sector were studied, along with a number of other models that considered the following additional variables:

- Predictions for more distant intervals t-30 and t+30
- Active and proposed flights (as separate variables)
- Look-ahead time (LAT)

For airport arrivals, equation (2) showed a substantial improvement (lower root mean square error (RMSE)) over equation (1). For equation (2) with short LAT (15 min – 1 hr), RMSE dropped from 3.45 to 2.83 (18% reduction); with medium LAT (1 – 2 hr), RMSE dropped from 3.75 to 3.12 (16.8% reduction).

For sector peak traffic volumes, the use of adjacent intervals also showed an improvement over current practice, but not as great as that for airports. For equation (2) with short LAT (15 min – 1 hr), RMSE dropped from 3.75 to 3.43 (8.5% reduction); with medium LAT (1 – 2 hr), RMSE dropped from 4.32 to 3.86 (10.6% reduction). Inclusion of the other variables (more distant intervals, proposed/active, LAT) did not add much value for either the airport or sector models.

Equation (2) was then tested on three days of data that were not part of the original calibration set. It continued to show improved accuracy. Furthermore, it showed reduced volatility, as measured by the number of times that the alert status changed. In short, this approach holds promise of providing predictions of airport and sector demand that are both more accurate and more stable.

Table of Contents

Executive Summary i

Table of Contents ii

Introduction 1

Section 1. Quality of Current Predictions 3

Section 2. New Models for Predicting the Number of Flights 8
    2.1 Adjacent Intervals 10
    2.2 Airport-Specific Constants...13
    2.3 Airport-Specific Coefficients...14
    2.4 Proposed and Active Flights 17
    2.5 Look-ahead Time 18
    2.6 Summary 21

Section 3. Testing Model 2 24
    3.1 Testing Model 2 on New Data..28

Section 4. Relationship between Flight Predictions and Monitor/Alert 29
    4.1 Model 2 Impact on Monitor/Alert..33

Section 5. Conclusion 35
    5.1 Summary of Study Results...35

Section 6. Next Steps 37

Section 7. References...37

Introduction

The Enhanced Traffic Management System (ETMS) predicts traffic demands that will be placed on airports, sectors, and fixes. The predictions are based mainly on flight plans, as well as radar tracking data for airborne flights. FAA traffic managers use these predictions along with other information such as weather forecasts to determine whether traffic flow management initiatives are needed during the next few hours.

Traffic Flow Management/Air Traffic Control (TFM/ATC) decision-making is mainly based on comparison of predicted traffic demand and available capacity at various National Airspace System (NAS) elements (airports, en route sectors, and fixes). For airports, demand is measured in aggregate number of aircraft per a specific time interval. ETMS considers traffic demand per 15-minute interval and relates it to 15-minute capacity. The aggregate demand count predictions are performed by predicting events for individual flights along the origin-destination routes (time and location) and then aggregating all the flights for a specific location and time interval. Even though it is well known that not all the predictions are 100 percent accurate, ETMS does not take into account uncertainty of the predictions and, hence, treats the predictions deterministically. Acknowledging existence of uncertainty in the demand predictions, and characterizing and quantifying the uncertainty, would make it possible to perform probabilistic demand forecasts that, in turn, would improve the ETMS prediction capabilities and TFM decision-making procedures.

There are several sources of errors in predictions of traffic demand. The source and magnitude of prediction errors depends on the status of flights. For 6 – 15-hour predictions, when mostly Official Airline Guide (OAG) data is available, there is uncertainty of whether the flight would fly at all, and if it would, its departure and arrival time as well as the route the flight would actually fly remain uncertain. As soon as the flight issued its flight plan, its route as well as proposed departure time may become more certain. In this case, overall uncertainty in both en-route events and arrival time decreased because of better (but still not ideal) knowledge of the flight's departure time and route. As soon as the flight becomes airborne, predictions of time and location of the flight along the route including arrival time become more accurate, more certain.

The purpose of this research is to examine the accuracy of predicted ETMS airport and sector counts, and to attempt to develop better prediction algorithms. The remainder of this report is in six sections.

- **Section 1   Quality of Current Predictions**
  Discussion of the accuracy of current flight predictions for a given location and 15-minute time interval.

- **Section 2   New Models for Predicting the Number of Flights**
  The development and calibration of new models for predicting the number of flights in a 15-minute time interval.

- **Section 3   Testing Model 2**
  Discussion of the testing of one promising new model.

- **Section 4   Relationship between Flight Predictions and Monitor/Alert**
  Discussion of the relationship between flight predictions and alerts, along with some performance metrics for alerts.
- **Section 5   Conclusion**
- **Section 6   Next Steps**
- **Section 7   References**

Section 1. Quality of Current Predictions

This section describes a review of the quality of current ETMS predictions of the number of flights in a 15-minute interval. This review was performed using 10 days of data from June and July 2005 for selected airports and sectors. For airports, the data includes the number of arrivals in a 15-minute interval. For sectors, the data includes the peak number of flights within a one-minute bucket of a 15-minute interval. ETMS predicts both of these quantities. In the absence of traffic flow management actions, the conceptual baseline model is

$A(t) = F(t,n) + \varepsilon$ where

$A(t)$ = Actual number of flights for 15-minute interval starting at time t (interval t)

$F(t,n)$ = Number of flights predicted at time n for interval t (The look-ahead time is t minus n.)

ε = A random error term (currently ignored by ETMS)

The error typically has a slightly asymmetric distribution. Figure 1 depicts the distribution of the error for sector ZBW02 for various look-ahead times (LAT). In this example, the average error ranges from 1 to 2 flights (corresponding to the first set of columns in Figure 5 on page 7) and the standard deviation of the error ranges from 2.5 to 6 flights (corresponding to the first set of columns for ZBW02 in Figure 3 on page 5).

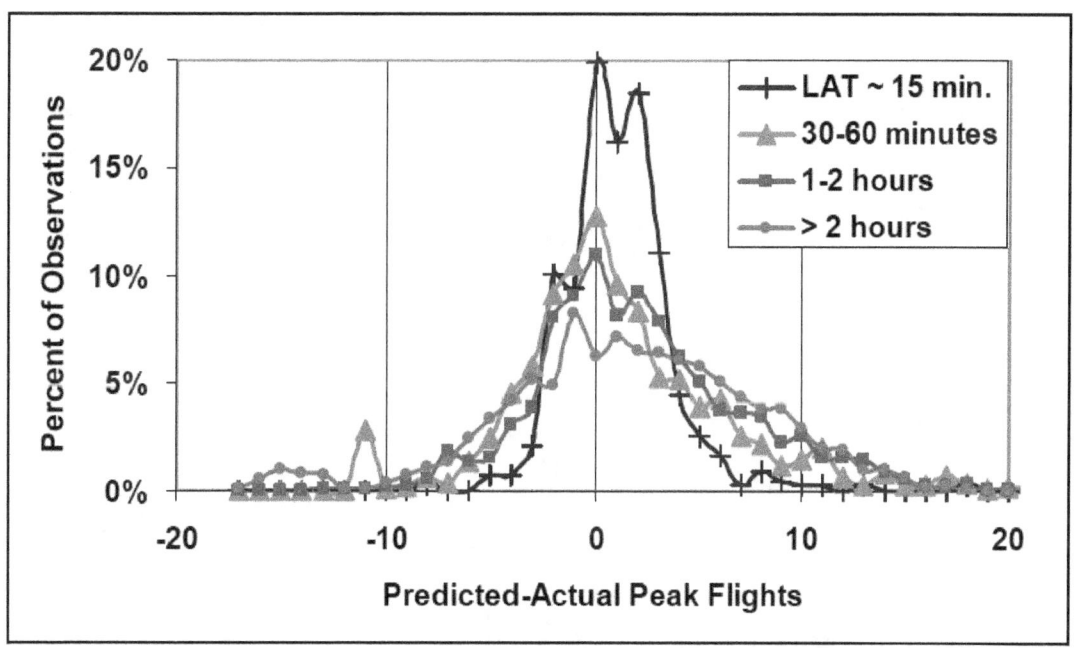

Figure 1 Histogram of Prediction Errors for ZBW02

Rather than showing separate histograms of the error for every airport and sector, the next few illustrations display the standard deviation and the bias to characterize both the shape and location of the histogram. The standard deviation is a measure of the dispersion of the error. If the error were to have a normal (bell-shaped) distribution, 68%

of the observations would fall within one standard deviation of the mean, and 95% of the observations would fall within two standard deviations. In Figure 1, the curve for LAT = 15 minutes has a standard deviation of 2.5 flights, while the curve for LAT > 2 hours has a standard deviation of 6 flights. If predictions were to be perfectly accurate, the standard deviation would be zero.

The bias is an indication of the location of the center of the distribution. In, the observations seem to be centered just to the right of zero (positive), and indeed, the bias for these curves ranges from 1 to 2 flights.

Figure 2 and Figure 3 show the standard deviation of error for 15-minute airport arrivals and sector peak counts, respectively, for several ranges of look-ahead times. Data was drawn from 10 days in June and July 2005. Nine airports and 13 sectors were examined. The time ranges used were as follows:

- Look-ahead time of 15 minutes
- Look-ahead time 30 minutes to 1 hour
- Look-ahead time more than 1 hour, and up to 2 hours
- Look-ahead time more than 2 hours

The numbers below each airport (Figure 2) are the average arrivals in 15 minutes. The numbers below each sector (Figure 3) are the average peak number of flights in 15 minutes.

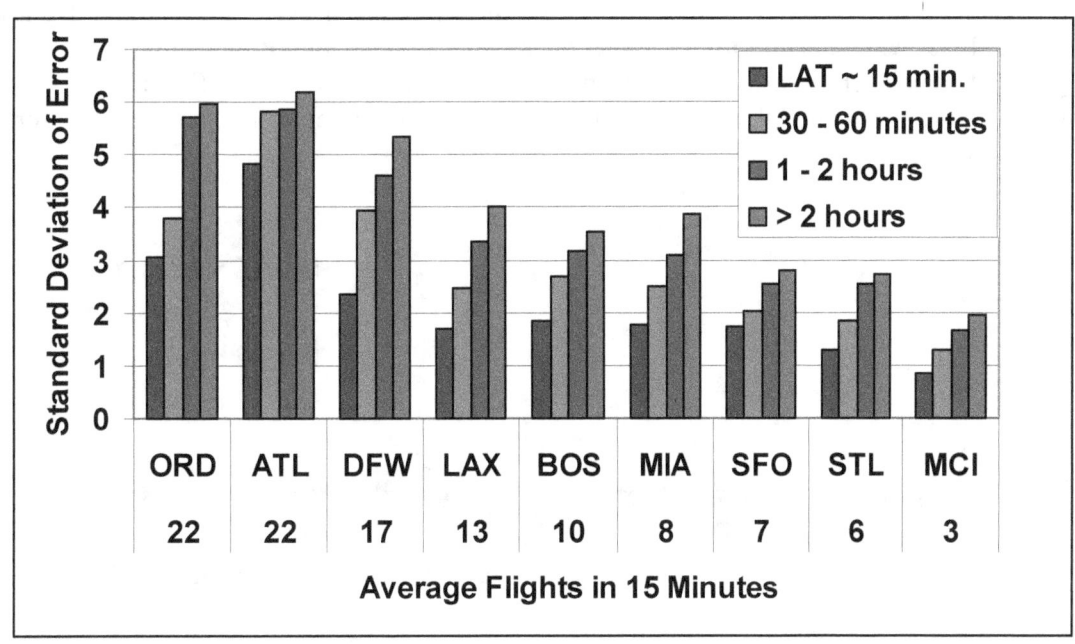

Figure 2 Standard Deviation of Error: Airports

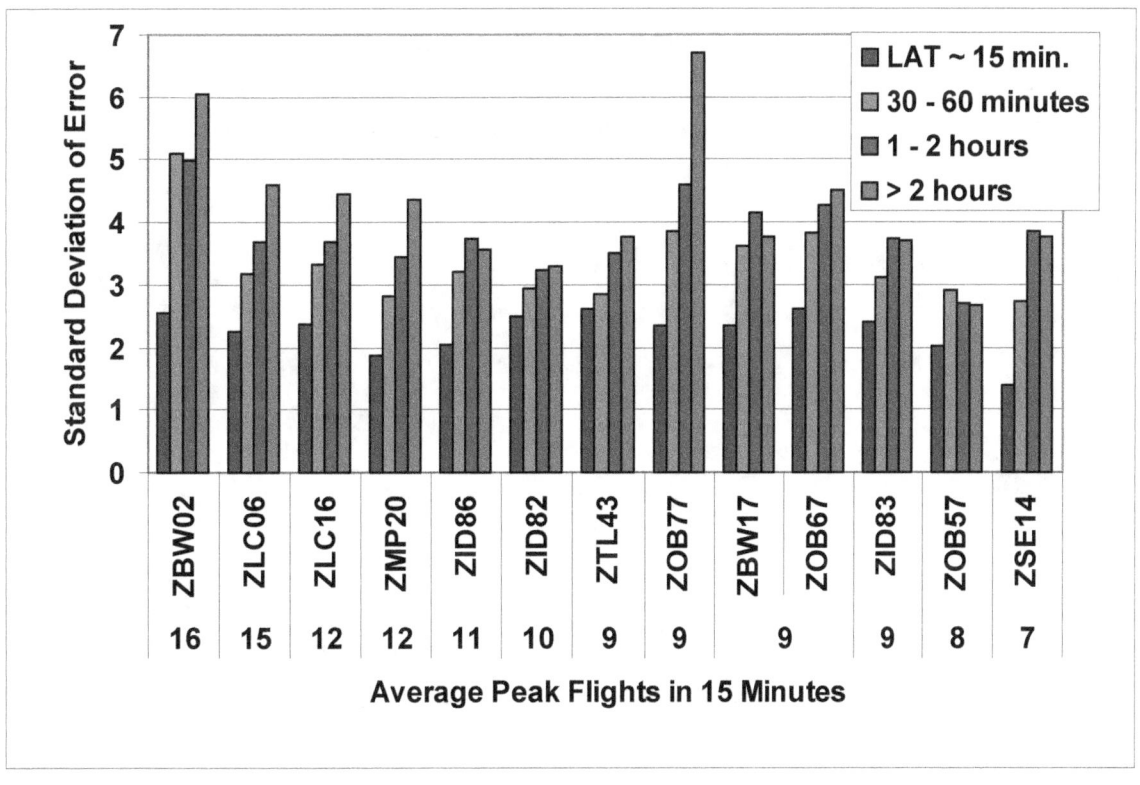

Figure 3 Standard Deviation of Error: Sectors

These results are not surprising. Both graphs show a larger dispersion (larger standard deviation of error) looking further into the future, while the airport graph (Figure 2) shows generally larger dispersion for the larger airports. Bias (*Average* (Predicted − Actual)) was generally small for airports (Figure 4), but somewhat larger for sectors (Figure 5). The high bias at ATL, ORD and DFW for short look-ahead times is unexpected, and requires further research.

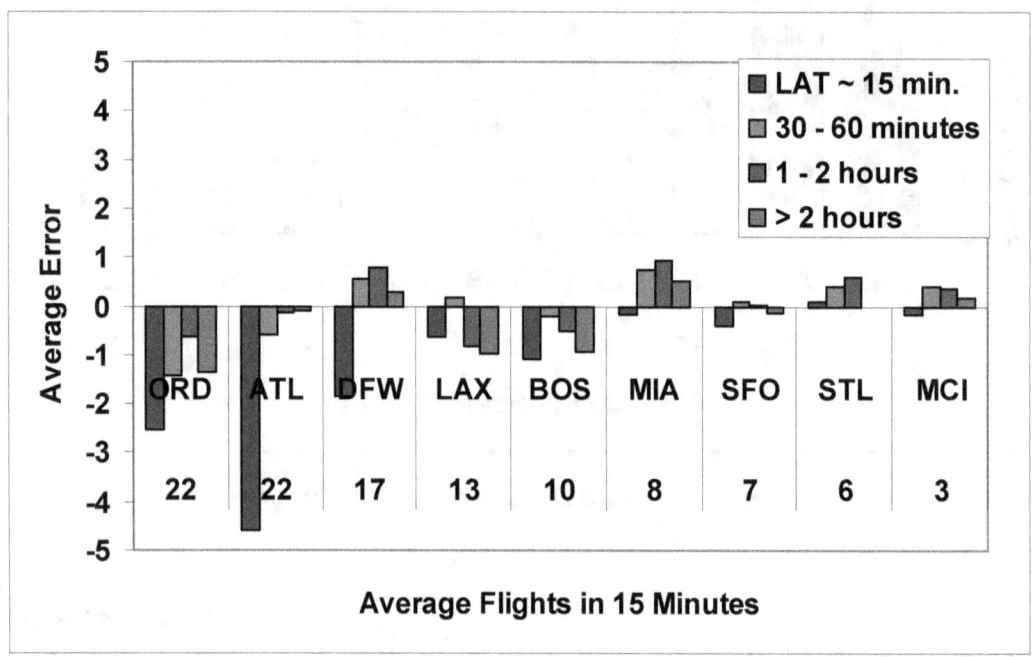

Figure 4 Prediction Bias: Airports

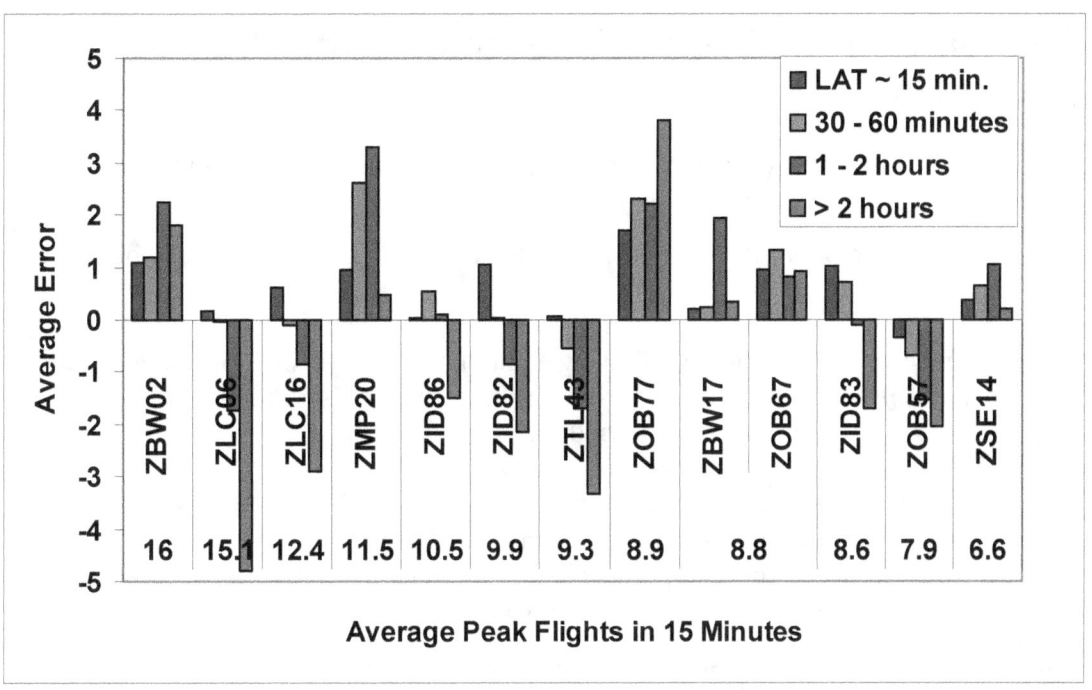

Figure 5 Prediction Bias: Sectors

Section 2. New Models for Predicting the Number of Flights

A variety of new models were tested using linear regression. The current model is

$A(t) = F(t,n) + \varepsilon$
(Actual number of flights at interval t equals Forecast made at time n for interval t plus some error)

Models were examined using the following additional variables, in various combinations:
 $F(t-30,n)$ = Forecast for two periods earlier (a period is 15 minutes)
 $F(t-15,n)$ = Forecast for one period earlier
 $F(t+15,n)$ = Forecast for one period later
 $F(t+30,n)$ = Forecast for two periods later
 $Active(t,n)$ = Active flights (airborne or landed flights)
 $Proposed(t,n)$ = Proposed flights (flights still on the ground at origin airports)
 $\sqrt{(t-n)}$ = Square Root of look-ahead time (LAT = $t-n$)

This research is intended to test which (if any) of these variables can contribute to improved flight predictions. Reasons for choosing these variables include the following:

- Given the uncertainty in estimated time of arrival (ETA) predictions, aircraft might migrate from one 15-minute interval to another. Therefore, it is useful to know the predictions for adjacent 15-minute intervals.

- Active flights are less likely to have substantial delays than proposed flights, therefore, active flights might be more likely to actually occur in the predicted time interval.

- Given the issue of pop-ups, predictions with a long look-ahead time might consistently underestimate the actual number of flights. Therefore, an adjustment based on look-ahead time might be appropriate.

Several sets of models were tested. The first (Models 2 and 3) looked at the impact of forecasts for adjacent intervals ($F(t-30,n)$, $F(t-15,n)$, etc.) on prediction accuracy for the interval of interest t. The second (Model 4) looked at the use of airport-specific constant terms in the regression model. The third (Model 5) considered proposed and active flights. The fourth (Model 6) looked at the square root of look-ahead time, and active/proposed flights. Finally, Models 7 and 8 combined some of the features of the previous models. In the models presented below, lower case letters a through h are coefficients, and k is a constant term. Because models 1 and 2 are later calibrated with two different data sets, they are subdivided into models 1a, 1b, 2a, and 2b. The only differences between the "a" and "b" variants are in the coefficients found by the regression model.

Model 0: $A(t) = bF(t,n) + \varepsilon$ (When $b=1$, this is the current ETMS model)
Model 1: (1a & 1b): $A(t) = bF(t,n) + k + \varepsilon$
Model 2: (2a & 2b): $A(t) = aF(t-15,n) + bF(t,n) + cF(t+15,n) + k + \varepsilon$
Model 3: $A(t) = gF(t-30,n) + aF(t-15,n) + bF(t,n) + cF(t+15,n) + hF(t+30,n) + k + \varepsilon$
Model 4: $A(t) = aF(t-15,n) + bF(t,n) + cF(t+15,n) + k(airport) + \varepsilon$
Model 5: $A(t) = dProposed(t,n) + eActive(t,n) + k + \varepsilon$
Model 6: $A(t) = dProposed(t,n) + eActive(t,n) + f\sqrt{(t-n)} + k + \varepsilon$
Model 7: $A(t) = aF(t-15,n) + bF(t,n) + cF(t+15,n) + dProposed(t,n) + f\sqrt{(t-n)} + k + \varepsilon$
Model 8: $A(t) = a_1Active(t-15,n) + a_2Proposed(t-15,n) + eActive(t,n) + dProposed(t,n) + c_2Active(t+15,n) + c_2Proposed(t+15,n) + k + \varepsilon$

2.1 Adjacent Intervals

It is well known that flights might migrate from one 15-minute interval to another due to a number of factors. Therefore, a prediction for one interval might provide some information for understanding what will happen in an adjacent interval. Analysis of accuracy of predictions for a 15-minute interval can be done, for example, by using a linear regression model with a random residual component and with variables that include predicted demand counts for the 15-minute interval of interest along with the preceding and following 15-minute intervals. Accordingly, models were developed that consider predictions on the adjacent intervals (Table 1 and Table 2). Both the single adjacent interval (t-15, t+15) and the two adjacent intervals (t-30, t-15, t+15, t+30) were tested.

The figure-of-merit used in Table 1 and subsequent tables is the Root Mean Square Error (RMSE). It is computed as $\sqrt{\frac{\sum\left(A(t)_i - \hat{A(t)}_i\right)^2}{n-2}}$, where n is the number of observations. It is similar to the standard deviation calculation except that the denominator is n-2, rather than n-1. Similar to the standard deviation used earlier (Figure 2), it is a shorthand for the dispersion of the error. Lower values indicate less error. One can expect that the majority of predictions will fall within one RMSE of the actual, and that (with normally distributed data) virtually all of the predictions will fall within 3 RMSEs of actual.

Because the regression combines a number of airports, the normally used figure-of-merit, R^2, is deceptively high. ETMS predicts high arrival rates for busy airports (such as ATL) and low arrival rates for less busy airports (such as MCI). When actual and predicted flights are plotted for several airports, the correlation will be very high, even though the quality of predictions might not be very high at an individual airport. For example, in model 1a for the shorter look-ahead time in Table 1, R^2 for all airports is 0.82, but if separate regressions were to be run for each airport, the R^2 for each regression would be considerably lower.

Table 1 Airport Arrival Models with Data from June 2005

	Model Number	0	1a	2a	3	0	1a	2a	3
	Data	Airports, 6/21/2005 – 6/25/2005							
		LAT = 15 min. – 1 hr				LAT = 1 hr – 2 hr			
	Number of Observations	3732				4322			
	Root Mean Square Error	3.45	3.43	2.83	2.75	3.75	3.71	3.12	3.04
	R-squared	0.81	0.82	0.87	0.88	0.78	0.79	0.85	0.86
Variable									
Constant	Coefficient	-	0.75	-0.5	-0.64	-	1.13	-0.20	-0.36
	Standard Error		0.11	0.1	0.09		0.11	0.10	0.07
$F(t-30,n)$	Coefficient	-	-	-	0.15	-	-	-	0.12
	Standard Error				0.011				0.011
$F(t-15,n)$	Coefficient	-	-	0.27	0.21	-	-	0.31	0.24
	Standard Error			0.011	0.011			0.011	0.012
$F(t,n)$	Coefficient	1.02	0.98	0.61	0.55	1.02	0.95	0.54	0.49
	Standard Error	0.004	0.008	0.011	0.011	0.004	0.008	0.012	0.012
$F(t+15,n)$	Coefficient	-	-	0.21	0.14	-	-	0.20	0.14
	Standard Error			0.011	0.012			0.011	0.012
$F(t+30,n)$	Coefficient	-	-	-	0.05	-	-	-	0.08
	Standard Error				0.011				0.011

For airport arrivals (Table 1), most of the coefficients are as expected. The largest coefficients are for $F(t,n)$, with smaller coefficients for adjacent intervals. For model 3, unexpectedly delayed flights contribute to a positive coefficient for $F(t-30,n)$ and $F(t-15,n)$, while early flights contribute to a positive coefficient for $F(t+15,n)$, and, to a lesser extent, for $F(t+30,n)$. Because, flights tend more often to be late than early, the coefficients for the prior intervals ($F(t-30,n)$, $F(t-15,n)$) are larger than those for future intervals. The non-zero constants are somewhat unexpected. Although the randomness of the data will lead the constants to be somewhat different than zero, these constants are significantly different. Model 2a shows a substantial improvement in predictive ability over models 0 and 1a, while model 3 shows only a slight improvement over model 2a.

Table 2 Sector Loading Models with Data from June 2005

Model Number		0	1a	2a	3	0	1a	2a	3
Data		\multicolumn{8}{c}{Sectors, 6/21/2005 – 6/25/2005}							
		\multicolumn{4}{c}{LAT=15 min. – 1 hr}	\multicolumn{4}{c}{LAT=1 hr – 2 hr}						
Number of Observations		\multicolumn{4}{c}{5326}	\multicolumn{4}{c}{6051}						
Root Mean Square Error		3.75	3.56	3.43	3.33	4.32	3.93	3.86	3.86
R-squared		0.61	0.65	0.67	0.69	0.49	0.58	0.59	0.60
Variable									
Constant	Coefficient	-	2.35	1.64	1.26	-	3.46	2.94	2.91
	Standard Error		0.10	0.10	0.10		0.10	0.18	0.11
$F(t-30,n)$	Coefficient	-	-	-	0.23	-	-	-	0.05
	Standard Error				0.013				0.014
$F(t-15,n)$	Coefficient	-	-	0.26	0.07	-	-	0.18	0.14
	Standard Error			0.014	0.017			0.014	0.019
$F(t,n)$	Coefficient	0.95	0.78	0.48	0.5	0.92	0.69	0.42	0.42
	Standard Error	0.004	.008	0.018	0.017	0.004	0.008	0.019	0.019
$F(t+15,n)$	Coefficient	-	-	0.011	0.09	-	-	0.14	0.16
	Standard Error			0.014	0.018			0.015	0.020
$F(t+30,n)$	Coefficient	-	-	-	-0.006	-	-	-	-0.03
	Standard Error				0.013				0.015

The situation for sectors (Table 2) is somewhat less clear. The model calculations are based on peak flights, so the impact of shifting flights earlier or later is less direct than it would be for a count of all flights. It appears that when the forecast of peak flights for a sector is unusually high, the actual number of peak flights is lower. This would help to explain the large constant terms and the coefficient that is substantially less than 1 in model 1a. It also helps to explain the coefficient that is less than 1 in Model 0. Even though ETMS may be correct in its predictions of the *total* number of flights in a sector over some time period, the dependent variable in this model is *peak* flights, not total flights. Therefore, it is possible for a model to consistently overpredict the peaks, while being correct in the total number of flights.

Comparing models 0 and 2a, where model 0 represents the current predictions, there appears to be a substantial benefit in considering the immediate adjacent intervals (t-15, t+15). The benefit of considering two adjacent intervals (model 3) is less clear. In particular, the large coefficient for $F(t-30,n)$ in model 3 appears to be an anomaly.

2.2 Airport-Specific Constants

The regressions have produced non-zero constant terms. The next model (model 4) tested the impact of estimating airport-specific constant terms. The data set used for this and subsequent models is larger than that presented in the earlier tables because it, in addition to the data from June 2005, also includes data from 5 days in July 2005. Models 1b and 2b are the same as models 1a and 2a, except that they were calibrated on the larger June/July dataset (Table 3).

Table 3 Models to Test Airport-Specific Constants

Model Number		1b	2b	4	1b	2b	4
Data		Airports from 6/21-6/25 and 7/13-7/17					
		LAT=30 min. – 1 hr			LAT=1 hr – 2 hr		
Number of Observations		11025	11025	11025	12480	12480	12480
Root Mean Square Error		3.24	2.75	2.67	3.79	3.33	3.12
R-squared		0.82	0.87	0.88	0.74	0.80	0.82
Variable							
Constant	Coefficient	0.73	-0.42	See Table 4	1.5	0.34	See Table 4
	Standard Error	0.058	0.05		0.06	0.06	
F(t-15,n)	Coefficient	-	0.25	0.19	-	0.31	0.19
	Standard Error		0.006	0.007		0.007	0.007
F(t,n)	Coefficient	0.94	0.61	0.55	0.87	0.51	0.40
	Standard Error	0.004	0.006	0.007	0.005	0.007	0.007
F(t+15,n)	Coefficient	-	0.19	0.13	-	0.16	0.05
	Standard Error		0.006	0.007		0.007	0.007

For model 4, the coefficients of the forecasts (F(t-15,n), etc) were forced to remain constant across all airports. Separate constant terms (k) were estimated for each airport, and are shown in Table 4.

Table 4 Airport-Specific Constants

	LAT=30 min. – 1 hr		LAT=1 hr – 2 hr	
	Coefficient k	Standard Error	**Coefficient k**	Standard Error
ATL	**3.454**	0.207	**8.008**	0.232
BOS	**1.442**	0.113	**3.809**	0.124
DFW	**1.709**	0.171	**5.562**	0.193
LAX	**1.419**	0.135	**5.119**	0.146
MCI	**0.110**	0.083	**0.954**	0.092
MIA	**0.354**	0.109	**2.375**	0.123
ORD	**4.048**	0.205	**8.599**	0.232
SFO	**0.758**	0.098	**2.579**	0.110
STL	**0.423**	0.095	**1.889**	0.107

It appears that the larger airports (such as ATL and ORD) have larger constant terms, especially for the 1-2 hour look ahead time. This indicates a tendency for flights to "fill in", e.g., when the ETMS deterministic forecast for a 15-minute period indicates few flights, the actual number of flights will be higher.

2.3 Airport-Specific Coefficients

Next, model 2b was calibrated on an airport-by-airport basis, with separate coefficients being computed for each airport.

Table 5 Airport-Specific Model Coefficients

	LAT=30 min. – 1 hr				LAT=1 hr – 2 hr			
	Constant	F(t-15,n)	F(t,n)	F(t+15,n)	Constant	F(t-15,n)	F(t,n)	F(t+15,n)
ATL	6.6	0.19	0.31	0.23	10.9	0.26	0.25	-0.01
BOS	0.9	0.19	0.58	0.16	2.7	0.17	0.40	0.18
DFW	4.3	0.09	0.62	0.01	6.2	0.07	0.48	0.05
LAX	-0.3	0.19	0.73	0.10	-0.3	0.35	0.62	0.13
MCI	0.1	0.05	0.70	0.11	0.2	0.12	0.58	0.13
MIA	1.0	0.11	0.58	0.11	1.5	0.16	0.45	0.12
ORD	3.7	0.28	0.55	0.06	15.2	0.15	0.25	-0.07
SFO	0.4	0.13	0.63	0.17	1.0	0.21	0.50	0.15
STL	-0.2	0.16	0.64	0.16	0.4	0.22	0.45	0.19

Figure 6 and Figure 7 depict the coefficients both for the non-airport-specific model 2 (labeled as "All Airports") and for the airport-specific models. The line is the constant term (with axis on the right), while the stacked bars are the F(t-15,n), F(t,n) and F(t+15,n) terms.

As expected, when the constant term is greater than zero, the stacked bars add up to less than 1, but when the constant term is close to zero, the stacked bars add up to a value near 1.

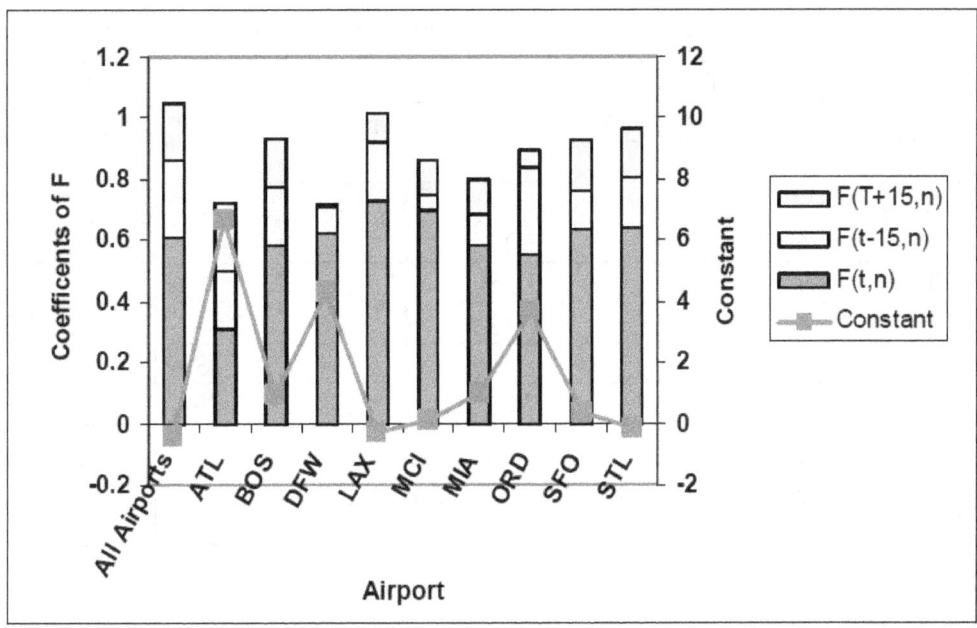

Figure 6 Coefficients for Airport-Specific Model, LAT=30 min. - 1 hour

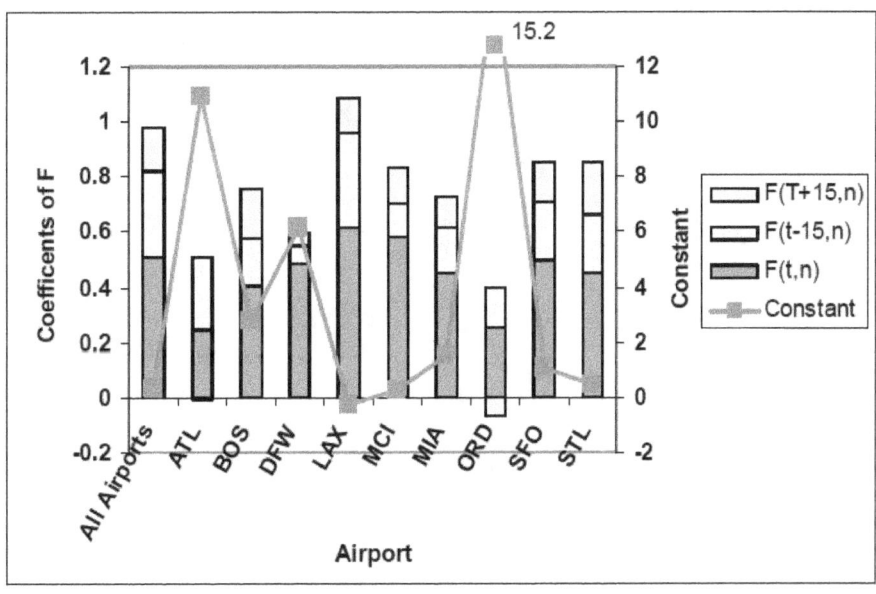

Figure 7 Coefficients for Airport-Specific Model, LAT = 1 – 2 hours

Standard deviation of error was also measured, for the current approach, model 2b (one model for all airports), and the airport-specific version of model 2b. While model 2b generally produced significantly better results than the current approach, the airport-specific model did not significantly outperform model 2b, except at the busiest airports, such as ATL and ORD (Table 6).

Table 6 Standard Deviation of Error: Airport-Specific Models

	LAT=30 min. – 1 hr			LAT=1 hr – 2 hr		
	Current	Model 2	Airport-Specific	Current	Model 2	Airport-Specific
All	3.27	2.75	2.58	3.91	3.33	3.01
ATL	5.81	4.44	4.10	5.86	4.93	4.49
BOS	2.69	2.26	2.25	3.15	2.55	2.50
DFW	3.94	3.73	3.55	4.62	4.24	4.06
LAX	2.47	2.24	2.16	3.36	2.86	2.82
MCI	1.29	1.21	1.13	1.64	1.43	1.37
MIA	2.52	2.22	2.15	3.08	2.49	2.43
ORD	3.81	3.23	3.15	5.70	4.78	3.92
SFO	2.02	1.73	1.67	2.56	2.08	2.04
STL	1.86	1.55	1.53	2.53	1.91	1.87

2.4 Proposed and Active Flights

Table 7 presents the impact of proposed and active flights on airport models. Differentiating between active and proposed flights (Model 5 versus model 1b) does not offer much improvement. Considering adjacent intervals (Models 2b and 8) does help. Note that F(t,n) = Proposed(t,n) + Active(t,n).

Table 7 Airport Models: Impact of Proposed / Active

Model Number		1b	5	2b	8	1b	5	2b	8
Data		Airports from 6/21-6/25 and 7/13-7/17							
		LAT=30 min. – 1 hr				LAT = 1 – 2 hr			
Number of Observations		11025				12480			
Root Mean Square Error		3.24	3.18	2.75	2.68	3.79	3.74	3.33	3.28
R-squared		0.82	0.82	0.87	0.87	0.74	0.75	0.80	0.81
Variable									
Constant	Coefficient	**0.73**	**0.81**	**-0.42**	**-0.32**	**1.5**	**1.5**	**0.34**	**0.30**
	Standard Error	0.058	0.06	0.05	0.052	0.06	0.06	0.06	0.059
F(t-15,n)	Coefficient			**0.25**				**0.31**	
	Standard Error			0.006				0.007	
F(t,n)	Coefficient	**0.94**		**0.61**		**0.87**		**0.51**	
	Standard Error	0.004		0.006		0.005		0.007	
F(t+15,n)	Coefficient			**0.19**				**0.16**	
	Standard Error			0.006				0.007	
Proposed (t,n)	Coefficient		**0.68**		**0.36**		**0.77**		**0.41**
	Standard Error		0.013		0.014		0.007		0.010
Active(t,n)	Coefficient		**0.99**		**0.65**		**0.98**		**0.62**
	Standard Error		0.005		0.007		0.008		0.011
Prop(t-15,n)	Coefficient				**0.29**				**0.33**
	Standard Error				0.018				0.010
Active(t-15,n)	Coefficient				**0.24**				**0.29**
	Standard Error				0.007				0.009
Prop(t+15,n)	Coefficient				**0.17**				**0.14**
	Standard Error				0.011				0.009
Active(t+15,n)	Coefficient				**0.20**				**0.17**
	Standard Error				0.007				0.011

In Table 8, regression results from sector data for June and July 2005, with a look-ahead time between 15 minutes and 2 hours, were examined. Similar to airports, differentiating between active and proposed flights (model 5 versus model 1b) does not offer much improvement.

Table 8 Sector Loading Models: Impact of Proposed/Active

	Model Number	1b	5
	Data	\multicolumn{2}{c}{Sectors from June/July 2005. LAT=15 min. – 2 hr}
	Number of Observations	33173	33173
	Root Mean Square Error	3.42	3.41
	R-squared	0.55	0.55
Variable			
Constant	Coefficient	3.1	3.2
	Standard Error	0.04	0.04
F(t-15,n)	Coefficient	-	-
	Standard Error		
F(t,n)	Coefficient	0.67	-
	Standard Error	0.003	
F(t+15,n)	Coefficient	-	-
	Standard Error		
Proposed(t,n)	Coefficient	-	0.63
	Standard Error		0.004
Active(t,n)	Coefficient	-	0.74
	Standard Error		0.006

2.5 Look-ahead Time

"**Probabilistic Congestion Management**" by Wanke et al [1] proposed using look-ahead time, proposed flights and active flights in a prediction model. Model 6, suggested by Wanke et al, splits Active and Proposed flights in the current period forecast, and considers look-ahead time. Signs of the coefficients are as expected. However, it does not appear that the $\sqrt{(t-n)}$ term adds much value, as the results from model 6 are similar to those from model 5 (Table 9).

Table 9 Airport Models: Impact of LAT

	Model Number	5	6	5	6
	Data	Airports from 6/21-6/25 and 7/13-7/17			
		LAT=30 min. – 1 hr		LAT = 1 – 2 hr	
	Number of Observations	11025	11025	12480	12480
	Root Mean Square Error	3.18	3.16	3.74	3.73
	R-squared	0.82	0.82	0.75	0.75
	Variable				
Constant	Coefficient	0.81	-1.60	1.49	-1.75
	Standard Error	0.06	0.251	0.064	0.43
Proposed (t,n)	Coefficient	0.68	0.61	0.77	0.75
	Standard Error	0.013	0.015	0.007	0.008
Active(t,n)	Coefficient	0.99	1.00	0.98	1.01
	Standard Error	0.005	0.005	0.008	0.009
$\sqrt{(t-n)}$	Coefficient	-	0.37	-	0.33
	Standard Error		0.038		0.043

Table 10 tests the impact of including $\sqrt{(t-n)}$ on the sector prediction models. Model 7 is a test of a combination of model 6 and the previously discussed adjacent period prediction model (model 2b).

Table 10 Sector Loading Models: Impact of LAT

Model Number		5	6	2b	7
Data		\multicolumn{4}{|c|}{Sectors from June/July 2005. LAT=15 min. – 2 hr}			
Number of Observations		33173	33173	33173	33173
Root Mean Square Error		3.41	3.40	3.32	3.30
R-squared		0.55	0.55	0.57	0.58
Variable					
Constant	Coefficient	3.2	1.8	2.4	1.3
	Standard Error	0.04	0.1	0.04	0.10
F(t-15,n)	Coefficient			0.23	0.23
	Standard Error			0.007	0.006
F(t,n)	Coefficient			0.39	0.49
	Standard Error			0.007	0.009
F(t+15,n)	Coefficient			0.10	0.10
	Standard Error			0.006	0.005
Proposed (t,n)	Coefficient	0.63	0.61		-0.15
	Standard Error	0.004	0.004		0.008
Active(t,n)	Coefficient	0.74	0.78		
	Standard Error	0.006	0.006		
√(t-n)	Coefficient		0.17		0.16
	Standard Error		0.011		0.011

For sectors, the signs and relative magnitude of the coefficients are as expected in model 6. However, the inclusion of √(t-n) did not result in a significant accuracy improvement over model 5.

Since the forecast flight variables (F(t-15,n), etc) are the sum of both active and proposed flights, there are two possible formulations for model 7:

$A(t) = 0.23F(t-15,n) + 0.49F(t,n) + 0.10F(t+15,n) + -0.15Proposed(t,n) + 0.16\sqrt{(t-n)} + 1.3 + \varepsilon$ (Table 5)

This is equivalent to

$A(t) = 0.23F(t-15,n) + 0.49(Active(t,n)+Proposed(t,n)) + 0.10F(t+15,n) + -0.15Proposed(t,n) + 0.16\sqrt{(t-n)} + 1.3 + \varepsilon$, or

$A(t) = 0.23F(t-15,n) + 0.49Active(t,n) + 0.34Proposed(t,n) + 0.10F(t+15,n) + 0.16\sqrt{(t-n)} + 1.3 + \varepsilon$

This last formulation may be compared with that in model 5.

2.6 Summary

Recall that the tested models were

Model 0: $A(t) = bF(t,n) + \varepsilon$ (When b=1, this is the current ETMS model)
Model 1a & 1b: $A(t) = bF(t,n) + k + \varepsilon$
Model 2a & 2b: $A(t) = aF(t-15,n) + bF(t,n) + cF(t+15,n) + k + \varepsilon$
Model 3: $A(t) = gF(t-30,n) + aF(t-15,n) + bF(t,n) + cF(t+15,n) + hF(t+30,n) + k + \varepsilon$
Model 4: $A(t) = aF(t-15,n) + bF(t,n) + cF(t+15,n) + k(airport) + \varepsilon$
Model 5: $A(t) = dProposed(t,n) + eActive(t,n) + k + \varepsilon$
Model 6: $A(t) = dProposed(t,n) + eActive(t,n) + f\sqrt{(t-n)} + k + \varepsilon$
Model 7: $A(t) = aF(t-15,n) + bF(t,n) + cF(t+15,n) + dProposed(t,n) + f\sqrt{(t-n)} + k + \varepsilon$
Model 8: $A(t) = a_1Active(t-15,n) + a_2Proposed(t-15,n) + eActive(t,n) + dProposed(t,n) + c_2Active(t+15,n) + c_2Proposed(t+15,n) + k + \varepsilon$

Figure 8 through Figure 12 compare the performance of the models, in terms Mean Square Error. Figure 8 and Figure 9 draw from Table 1, Table 3, Table 7 and Table 9. Figure 10 and Figure 11 draw from Table 2. Figure 12 draws from Table 8 and Table 10.

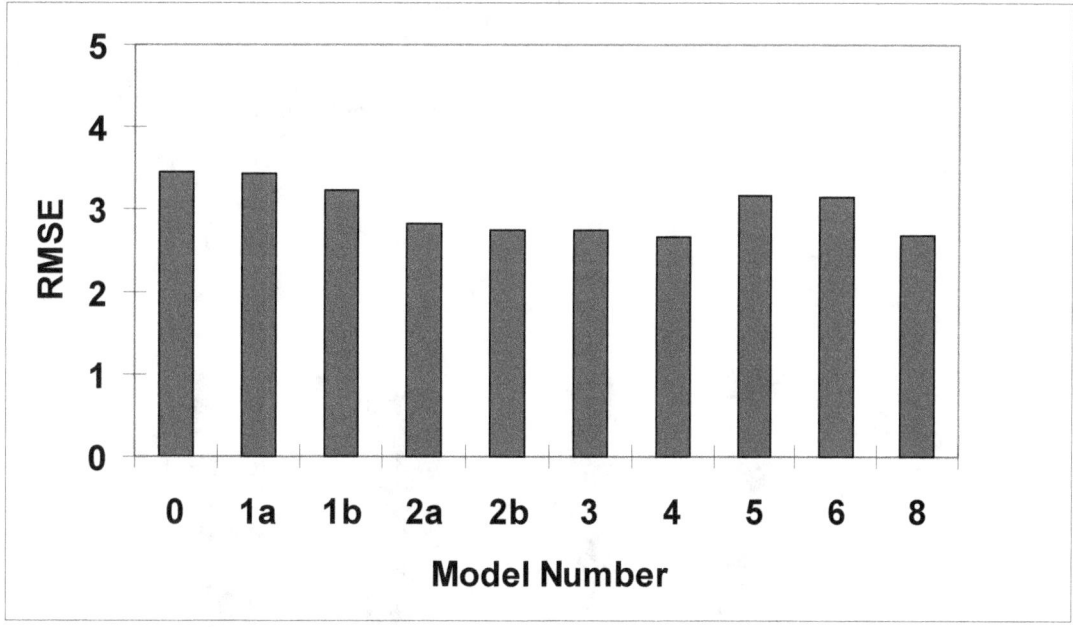

Figure 8 Airport Models: Short Look-ahead

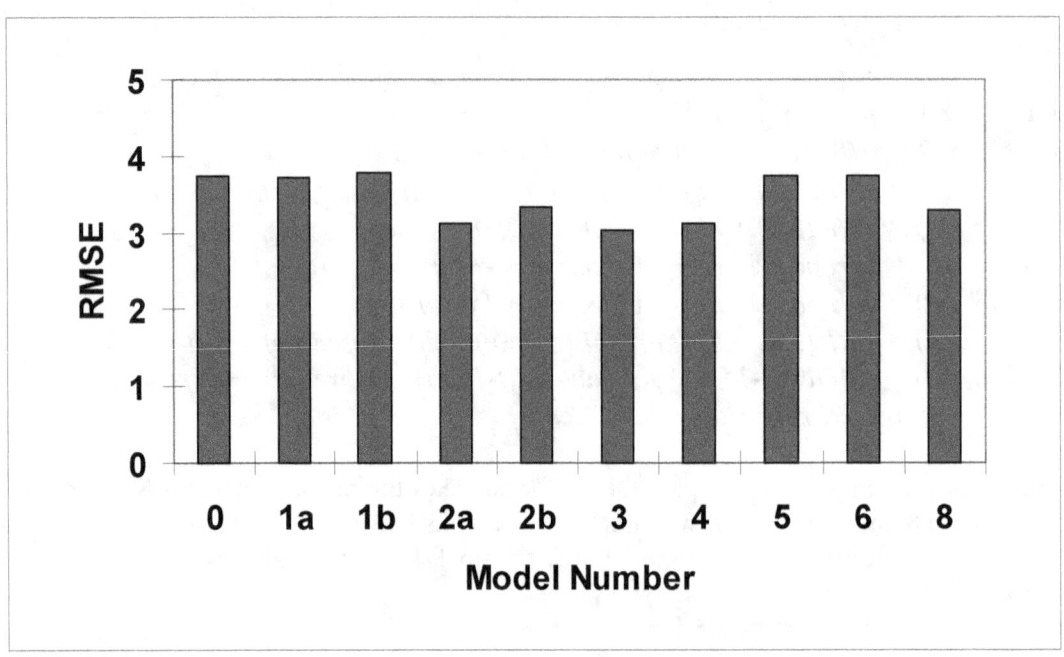

Figure 9 Airport Models: Medium Look-Ahead

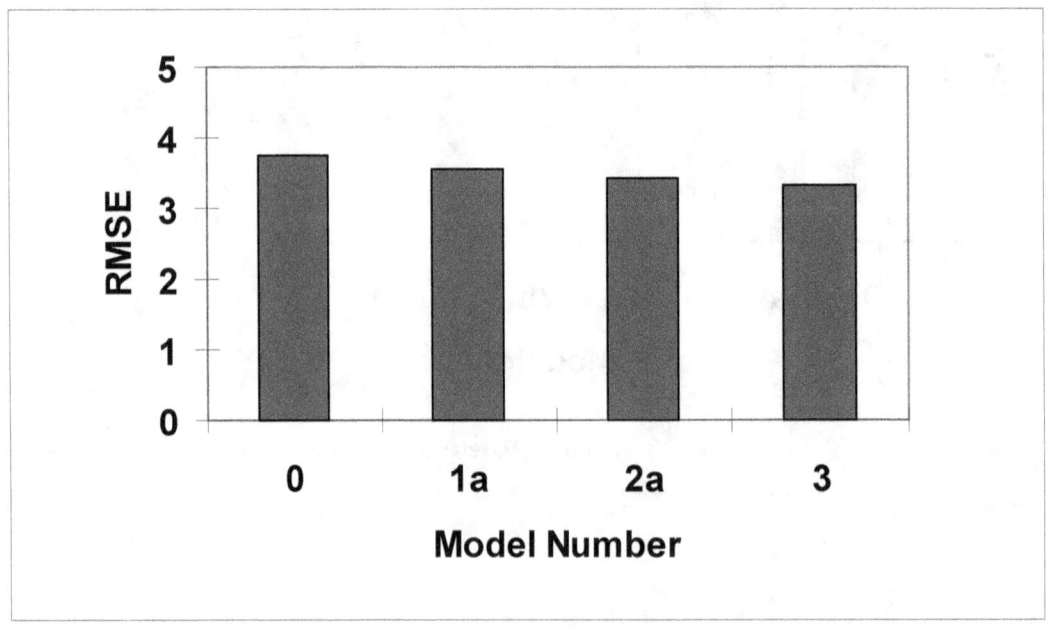

Figure 10 Sector Models: Short Look-Ahead

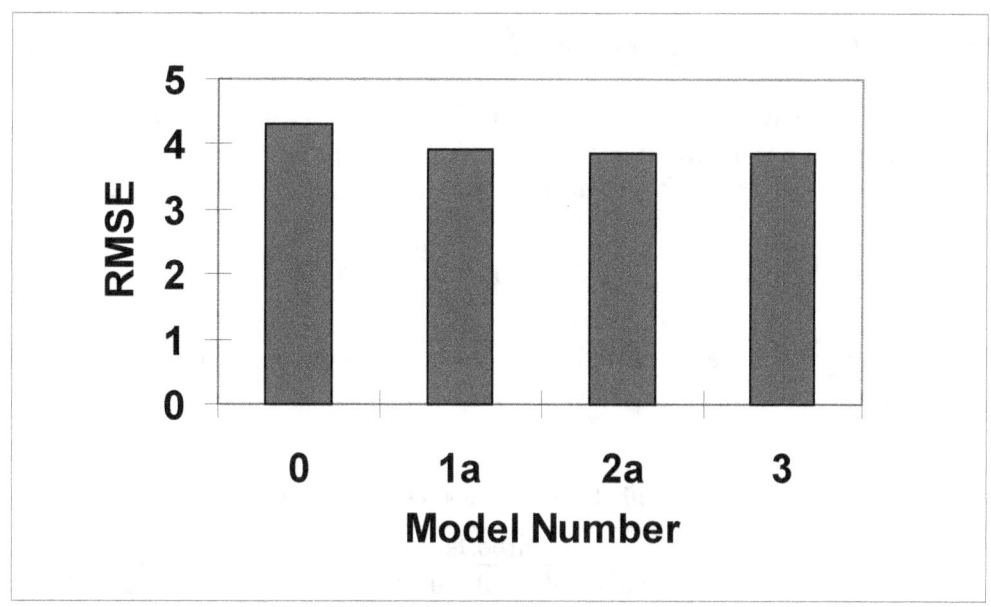

Figure 11 Sector Models: Medium Look-ahead

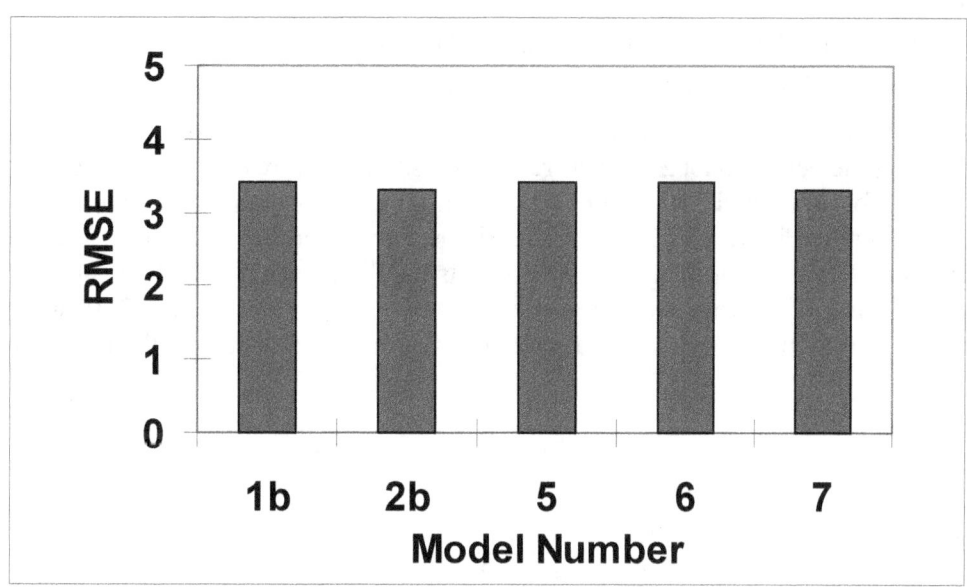

Figure 12 Sector Models: Mixed Look-ahead

Section 3. Testing Model 2

Given the above regression results, it appears that model 2 (with a single set of coefficients instead of airport-specific coefficients) offers the best combination of both simplicity and improved results. Therefore, it was further tested using the following formula:

$$F_{new}(t,n) = aF(t-15,n) + bF(t,n) + cF(t+15,n) + k$$

After examining the model 2 coefficients in the tables in the previous section, the following values were chosen for a, b, c and k (Table 11).

Table 11 Coefficients for New Model

	Airports			Sectors	
Look-ahead time:	~ 15 min.	30 – 60 min.	> 1 hour	15 – 60 min.	> 1 hour
k	0	0	0	2	3.5
a	0.3	0.3	0.3	0.25	0.2
b	0.6	0.55	0.5	0.45	0.4
c	0.1	0.15	0.2	0.05	0.1

Recall that the model currently used in ETMS has the form $F_{current}(t,n) = F(t,n)$, with $k=a=c=0$ and $b=1$.

$F_{new}(t,n)$ is model 2 with coefficients from Table 11. $F_{current}(t,n)$ is the current ETMS model. Comparing the new (model 2) and current models, biases were similar, but standard deviations for individual airport and sector predictions were somewhat lower for the new model. Compare Figure 13 with Figure 2, Figure 14 with Figure 3, Figure 15 with Figure 4, and Figure 16 with Figure 5. Figure 17 presents the difference between Figure 13 and Figure 2, while Figure 18 presents the difference between Figure 14 and Figure 3. Positive values in Figure 17 and Figure 18 indicate that model 2 is an improvement over the current model.

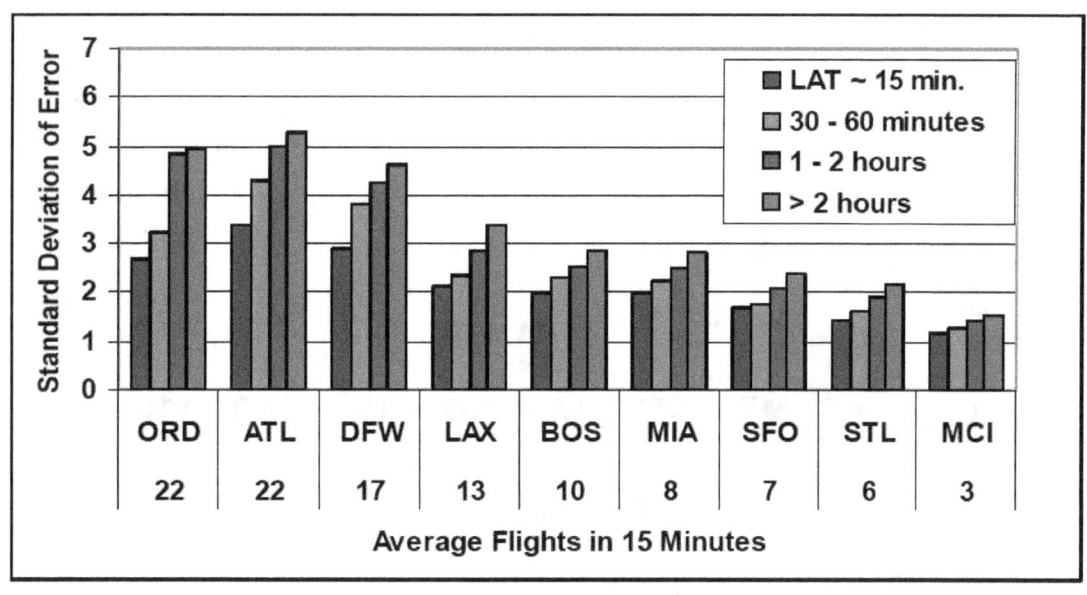

Figure 13 New Prediction Standard Deviations: Airport

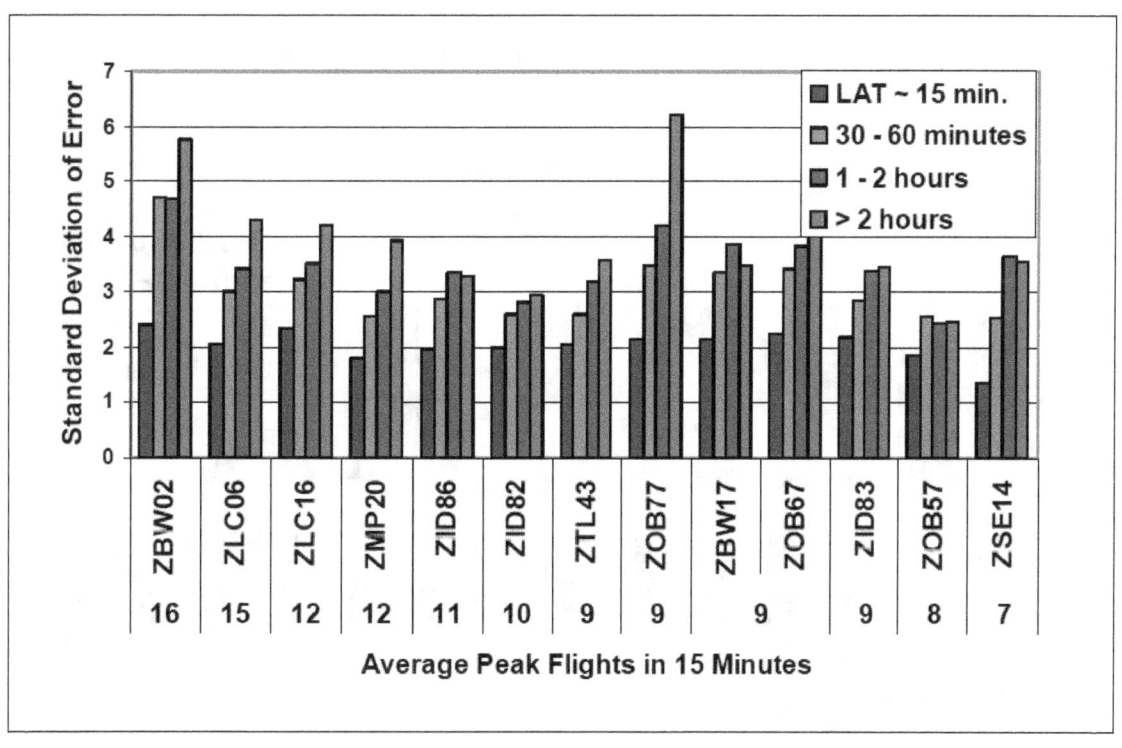

Figure 14 New Prediction Standard Deviations: Sectors

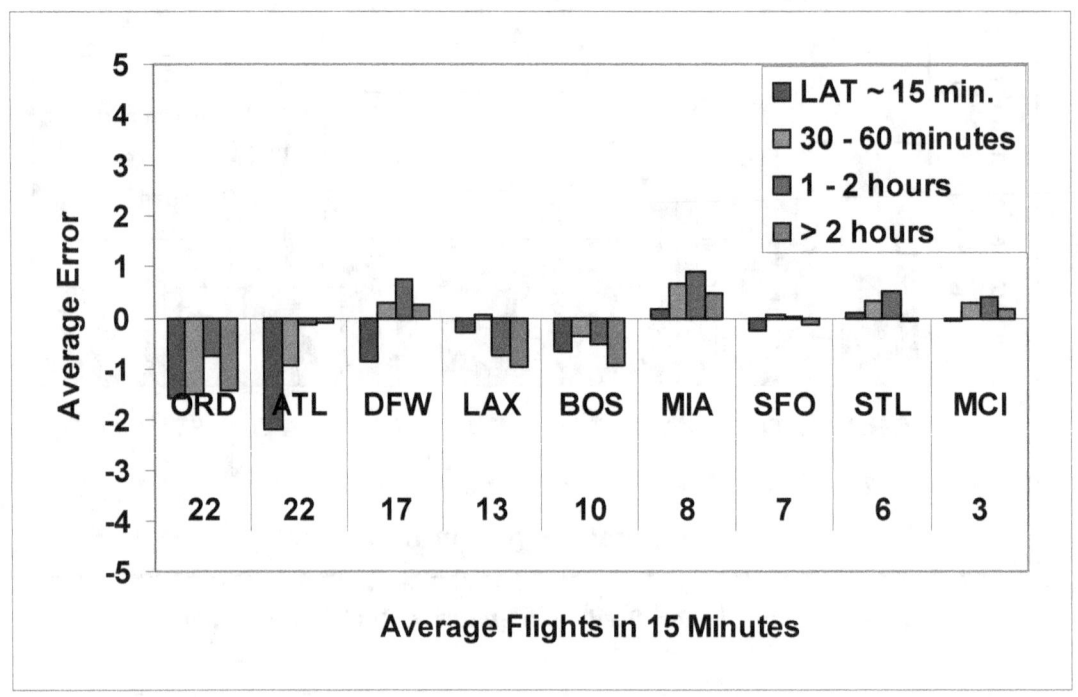

Figure 15 New Prediction Bias: Airports

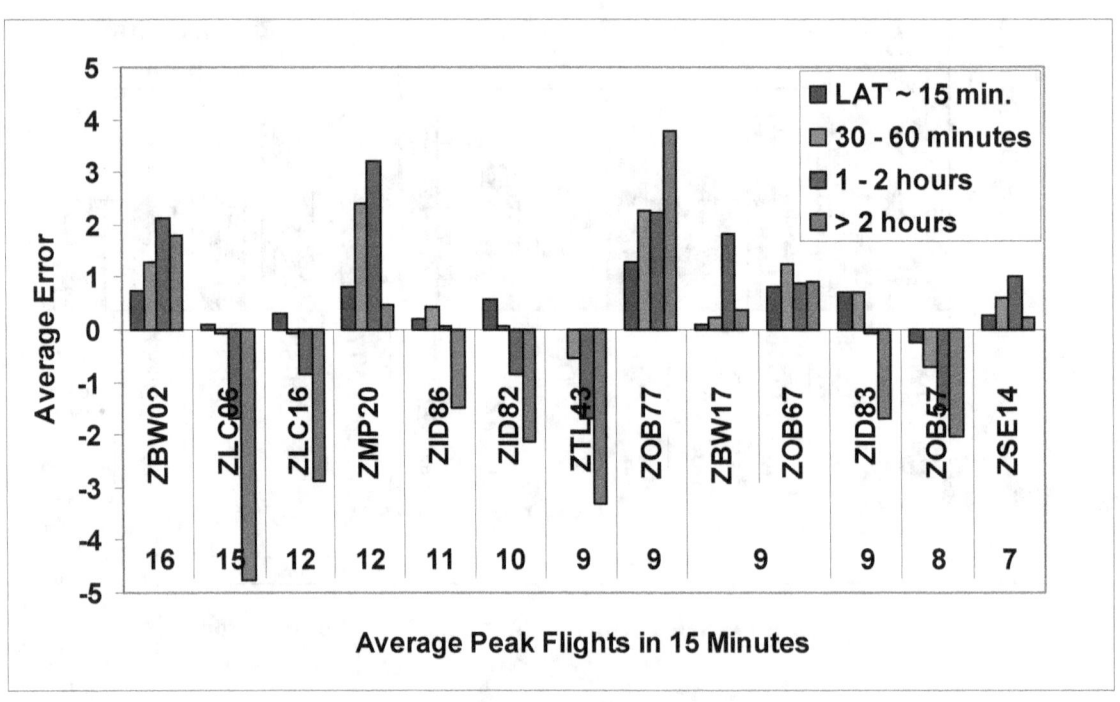

Figure 16 New Prediction Bias: Sectors

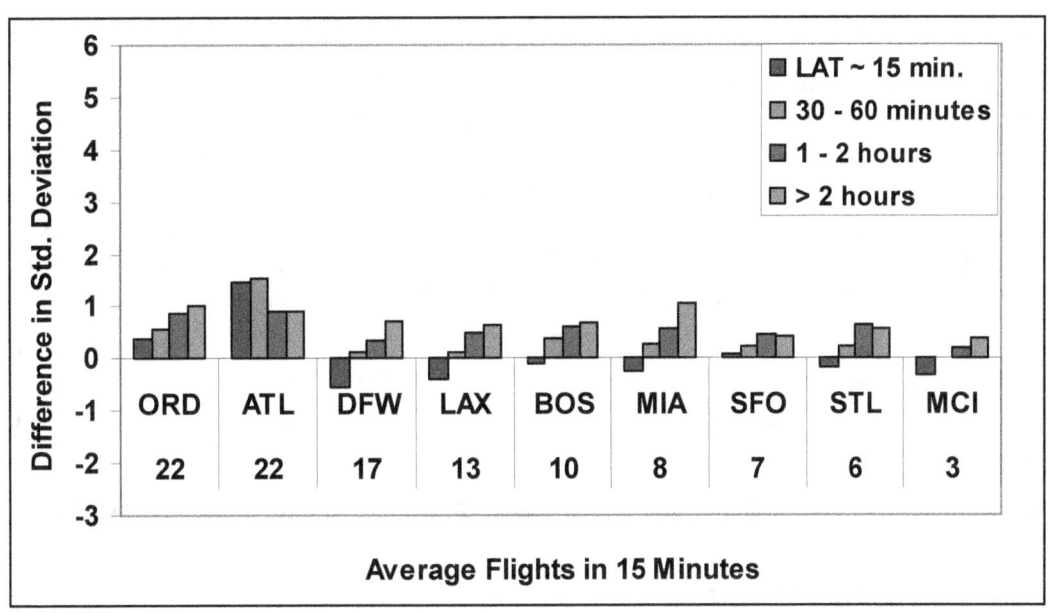

Figure 17 Difference (Current – New) in Prediction Error: Airports

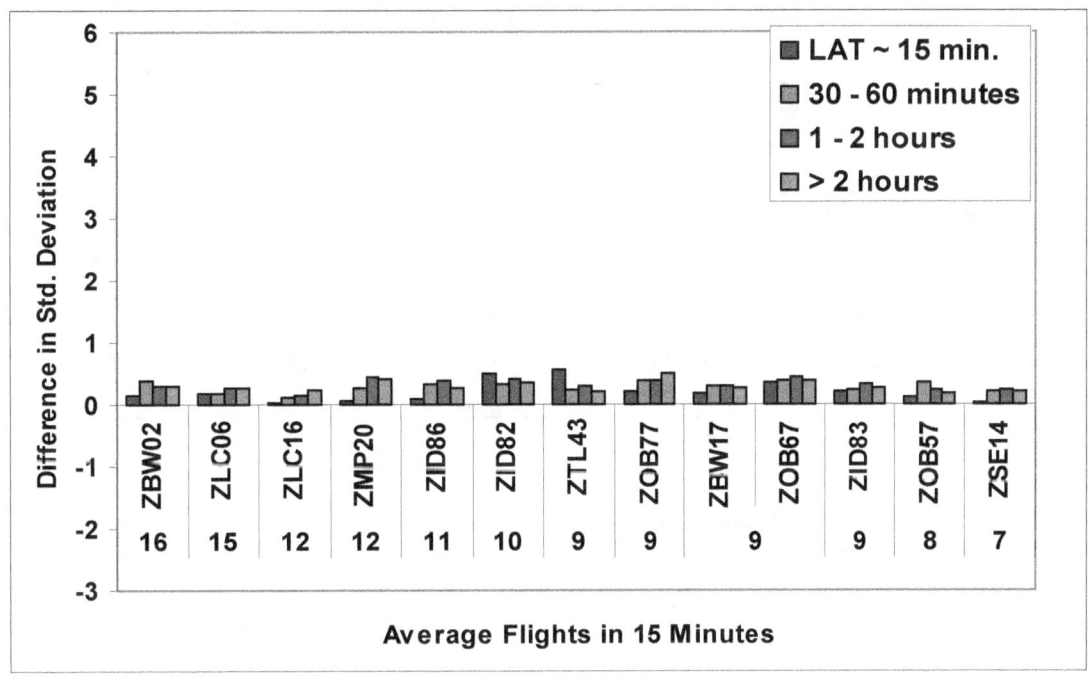

Figure 18 Difference (Current – New) in Prediction Error: Sectors

For airports, the new model showed a lower standard deviation of prediction error for time bins 2 through 4 (Figure 17). For sectors (Figure 18) the difference was very small, but still consistently showed a lower standard deviation of prediction error for the new model. Bias was better for large airports (ORD, ATL) in the airport model, but similar for the smaller airports and for the sector model.

3.1 Testing Model 2 on New Data

The accuracy and volatility of predictions from model 2 were tested using 3 days of data that were not in the calibration set. The data included some 24,000 predictions for various airport, time of prediction (n), and event time (t) combinations, and some 33,000 predictions for various sector, time of prediction (n), and event time (t) combinations. The three days, all in 2005, were Friday July 22, Saturday July 23, and Wednesday July 27. The Friday was roughly an "average" day (as judged by the number of ATCSCC advisories issued), Saturday was a "quiet" day, and the Wednesday was a day when a large number of advisories were issued.

As mentioned earlier, one obvious measure of error is $F(t,n)-A(t)$, which can measure both the mean (to assess bias in the forecasts) and the standard deviation (to assess dispersion) of a number of observations.

A measure of volatility is the number of times a prediction switches between exceeding and not exceeding capacity.

In all of these measures, it is more desirable for the value to be closer to zero. Table 12 shows the values for these measures for airports and Table 13 shows the values for sectors. In all cases, they are as good or better in the new model.

Table 12 Airports: Current Model vs. Model 2

		Current	Model 2
Standard Deviation of Error	7/22	5.5	5.2
	7/23	3.2	2.6
	7/27	6.8	6.5
Average Error (bias)	7/22	-1.7	-1.7
	7/23	0.2	0.2
	7/27	-2.3	-2.3
Number of times predicted flights crosses the capacity threshold	7/22	1079	948
	7/23	480	265
	7/27	789	562

Table 13 Sectors: Current Model vs. Model 2

		Current	Model 2
Standard Deviation of Error	7/22	4.5	3.6
	7/23	4.6	3.7
	7/27	5.5	4.5
Average Error (bias)	7/22	-0.7	-0.6
	7/23	-0.2	0.1
	7/27	-2.2	-1.3
Number of times predicted flights crosses the capacity threshold	7/22	885	880
	7/23	795	640
	7/27	566	511

Section 4. Relationship between Flight Predictions and Monitor/Alert

This section covers the relationship between predictions for individual 15-minute intervals and Monitor/Alert predictions over all the intervals in a user-chosen planning. This discussion assumes that the planning horizon is two hours.

ETMS Monitor/Alert functionality for airports compares predicted traffic demand for each 15-minute interval with airport capacity (separately for arrivals and departures). If demand exceeds capacity of a 15-minute interval, Monitor/Alert identifies the interval as congested and issues an alert warning (red or yellow depending on the status of the flights).

A "good" prediction for Monitor/Alert has two characteristics

- It correctly indicates when an airport or sector is likely to be overloaded or not overloaded
- It does so without much "flicker", i.e. once a sector or airport is alerted, it should stay alerted, until air traffic control intervention occurs.

To illustrate these concepts, consider the following example. An airport[1] has an arrival capacity of 20 flights in 15 minutes. In the absence of TFM intervention, the actual number of arrivals builds from zero at 4 AM to approximately 24 flights/15 minutes between 8 and 8:30 AM (Figure 19).[2]

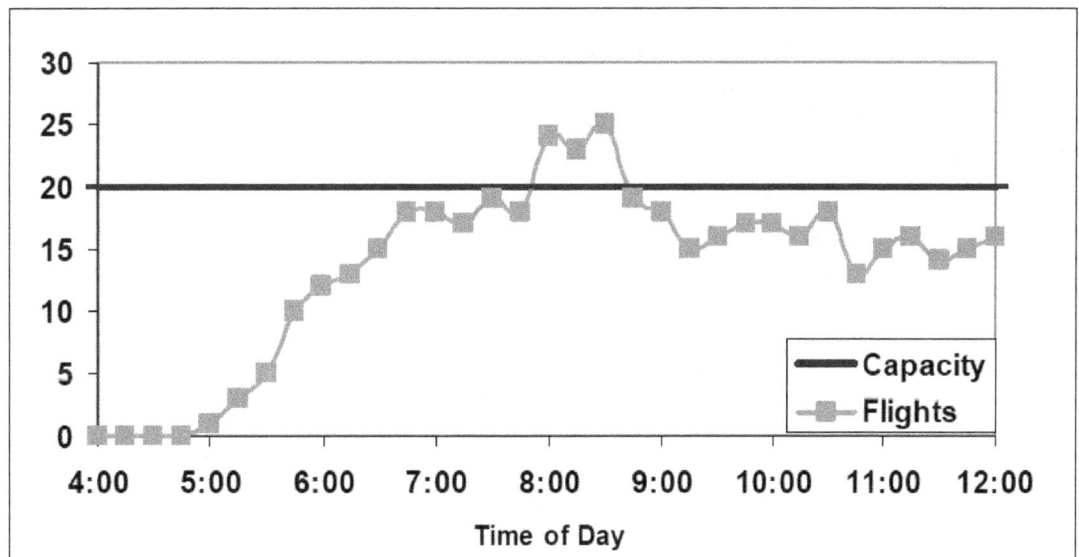

Figure 19 Airport Arrivals by 15-minute Interval

---

[1] The logic of this example also applies to sectors, except that for sectors the measure of interest is peak number of flights in a sector, rather than total flights in 15 minutes.

[2] In this example, the actual number of arrivals sometimes exceeds the stated capacity. This appeared to occur fairly frequently (with the number of arrivals in the current 15-minute interval exceeding capacity), suggesting that the published capacity may be somewhat less than the actual capacity for observed traffic.

ETMS shows two types of alerts:

1. Alert status for an individual 15-minute interval: an interval for an airport or sector is alerted if demand (number of arrivals for airports, or peak number of flights for sectors) exceeds capacity.
2. Alert status for airports/sectors as a whole: an airport or sector is alerted if demand exceeds capacity during at least one 15-minute interval in the planning horizon, where the planning horizon is typically about 2 hours.

It is desirable that the prediction of arrivals for a 15-minute interval be both accurate and stable. Accuracy means that the number of flights for a 15-minute interval is accurately predicted at least several hours in advance. Stability means that the prediction does not change much over the hours between the first prediction and the actual event.

The relationship between arrivals and Monitor/Alert is as follows:

- There is an established planning horizon (for example, assume it is two hours)
- There is a set of predictions of arrivals (one for each 15-minute interval) for intervals occurring between now and the planning horizon. For a planning horizon of 2 hours, 8 predictions would be considered.
- If any one of the predictions exceeds the capacity, the airport is alerted. Note that this is equivalent to saying that if the maximum of the predictions exceeds the capacity, the airport is alerted. Therefore, to determine whether an alert should occur, the maximum of a set of predictions is plotted (See Figure 20 on page 31 and Figure 21 on page 32).

In the example of Figure 19, there should be an alert between 6 AM (2 hours before 8 AM, when the number of flights first exceeds the capacity) and 8:30 AM (the end of the period where the number of flights exceeds capacity (Figure 20)). Figure 20 plots the maximum of a set of forecasts occurring during the look-ahead period:

$$M/A\ Value = \mathrm{Max}(F(n,n), F(n+15\ minutes, n)...F(n+120\ minutes, n))$$

A NAS element is alerted when the *M/A Value* exceeds the alert threshold.

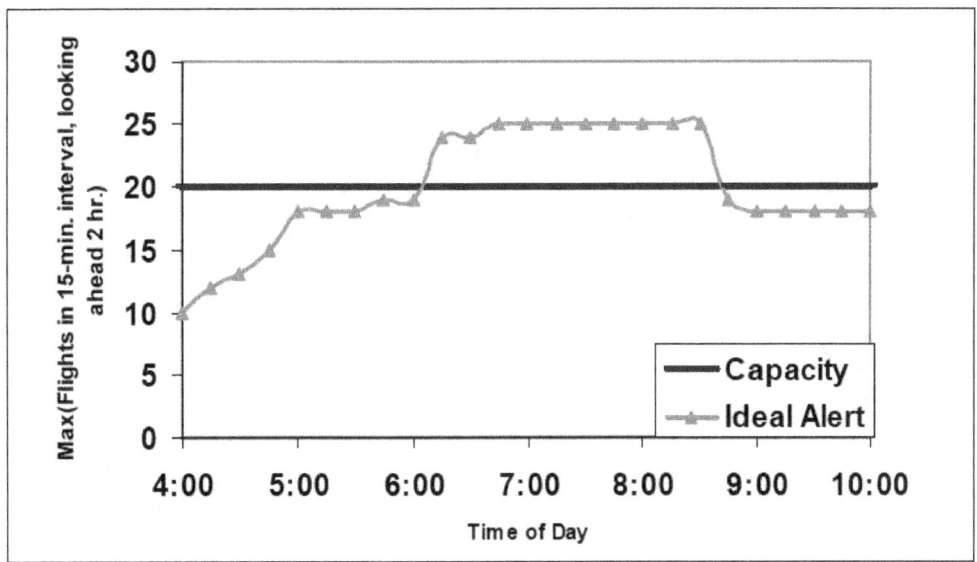

Figure 20 Ideal Alert

There are several ways that alerts can fail to meet this ideal (Figure 21):

- An alert may not occur, even though the actual number of flights exceeds capacity (Fails to Alert). In the example (Figure 21) the predicted maximum fails to exceed capacity between 6 AM and 8:30 AM, although the actual maximum does.
- An alert may occur even though the number of flights is under capacity (False Alert). In the example (Figure 21), the predicted maximum remains above capacity at 9 AM, even though the actual maximum has dropped.
- Finally, the predicted number of flights may fluctuate wildly, leading to alerts being switched on and off (Volatility). In the example, the predicted maximum number of flights crosses the capacity line multiple times, indicating that alerts will be switched on and off.

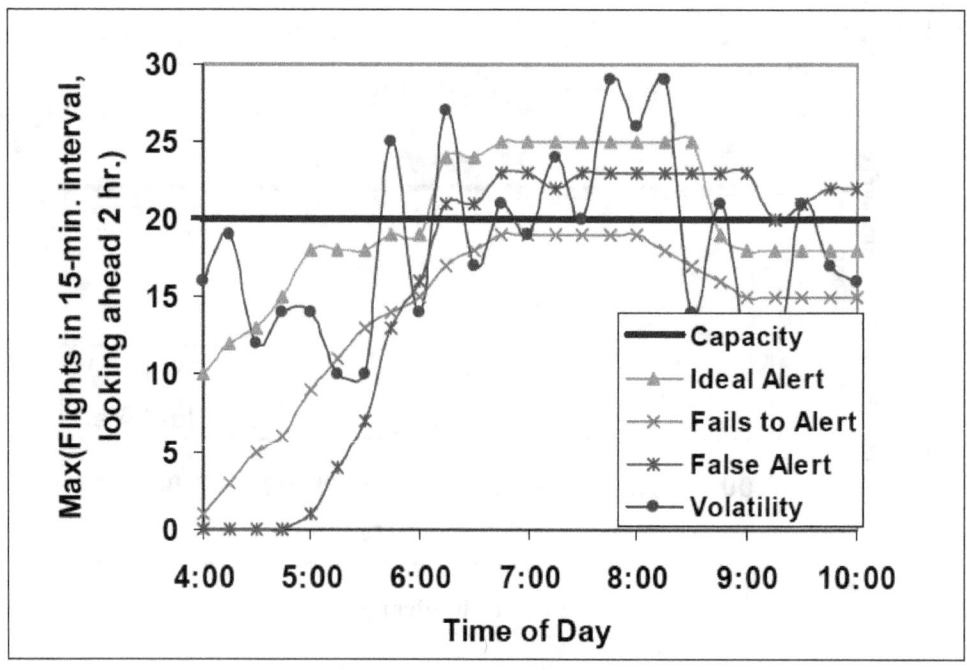

Figure 21 Types of Errors in Flight Predictions

With the above discussion, there are a number of measures for assessing the quality of predictions that relate directly to the Monitor/Alert predictions. As such, they need an assumption on look-ahead times. The measures include

- Number of alerts
- Number of times alerts are switched on (a low number is good).

Unfortunately, given that TFM actions are likely to be taken, it is difficult to assess whether an alert *should* have been switched on. For example if a sector is alerted, but in the end does not see flights exceeding capacity, was this because the alert was incorrect, or because a TFM action was taken?

Table 14 shows the values of some of these measures, for the example in Figure 21.

Table 14 Monitor/Alert Measures

	Scenario (from Figure 21)			
Measure	Ideal Alert	Fails to Alert	False Alert	Volatility
Alerts	10	0	16	10
Number of times alerts are switched on	1	0	1	6

With a look-ahead of 2.5 hours, Monitor/Alert statistics for June and July data are as follows. Table 15 compares what actually occurred to what would have occurred had Monitor/Alert been perfectly accurate (assuming no TFM actions).

Table 15 Monitor/Alert Measures: June/July Data

Measure	Airport Arrivals		Sector Peaks	
	Ideal Alert	Actual	Ideal Alert	Actual
Alerts	881	956	1091	1325
Number of times alerts are switched on	22	108	73	242

4.1 Model 2 Impact on Monitor/Alert

A sector or airport is alerted if any of the 15-minute flight forecasts during the look-ahead time exceeds a threshold. Recall that this is equivalent to saying that a sector or airport is alerted if the maximum of the 15-minute flight forecasts exceeds the threshold. Therefore,

M/A Value = Max$(F(n,n), F(n+15\ minutes, n), F(n+30\ minutes, n) \ldots F(n+120\ minutes, n))$

A new model (F_{new}) was developed based on model 2, using coefficients in Table 11:

Model 2 M/A Value = Max$(F_{new}(n,n), F_{new}(n+15\ minutes, n) \ldots F_{new}(n+120\ minutes, n))$

The new model was tested using three days of data that were not part of the calibration set: 7/22, 7/23 and 7/27/2005. Table 16, Table 17, and Table 18 compare both the number of alerts, and the number of times alerts are switched on for the current model and model 2.

Table 16 Monitor/Alert Measures: July 22, 2005

Measure	Airport Arrivals		Sector Peaks	
	Current Model	Model 2	Current Model	Model 2
Alerts	103	76	171	84
Number of times alerts are switched on	18	8	30	15

Table 17 Monitor/Alert Measures: July 23, 2005

Measure	Airport Arrivals		Sector Peaks	
	Current Model	Model 2	Current Model	Model 2
Alerts	98	80	166	74
Number of times alerts are switched on	10	3	30	17

Table 18 Monitor/Alert Measures: July 27, 2005

Measure	Airport Arrivals		Sector Peaks	
	Current Model	Model 2	Current Model	Model 2
Alerts	104	82	100	21
Number of times alerts are switched on	11	2	23	6

Since it is unknown whether TFM actions were taken in response to the alerts, comparing the alerts to the "Ideal" is not helpful. If a sector or airport is alerted, but the number of flights ends up being below the alert threshold in reality, it is unknown whether this occurred due to an TFM action (in which case the alert was correct) or whether it occurred in the absence of a TFM action (in which case the alert was incorrect). Therefore, a judgment cannot yet be made as to which model is "better."

The major difference is that model 2 produces fewer alerts than the current model. This is not surprising. Recall that the M/A Value is based on the maximum of a set of forecasts. If there is less volatility in the forecasts, the maximum of the forecasts will tend to be lower than the previous maximum. Therefore, an airport or sector will be less likely to be alerted.

Section 5. Conclusion

ETMS currently makes its predictions based on deterministic projections of traffic. Predictions based on this model can be improved by including in the calculation a factor for uncertainty. For example, if flights are frequently and unexpectedly delayed, the model can include a deterministic projection for a prior period (t-1) along with the current projection for period t to improve the prediction of what will actually happen at period t.

Regression analysis was used to develop several new models for predicting the number of flights in a 15-minute interval. The models are also used for predicting airport arrivals and for predicting the peak number of flights in a sector. These models considered the following new variables:

- Predictions for adjacent intervals
- Active and proposed flights (as separate variables)
- Look-ahead time (LAT)

5.1 Summary of Study Results

The results of the study can be summarized as follows:

1. Statistical data analysis of historical ETMS data shows that the amount of uncertainty in traffic demand predictions at arrival airports and en route sectors depends on both the specific NAS element and look-ahead time. The analyzed data covered 10 days: five days in June 2005 and five days in July 2005. Characteristics of the error include
   - Bias (average value different from zero). The sign and magnitude of the bias varied by airport/sector and look-ahead time. In most cases, the bias was between –2 and +2 flights per 15-minute time bucket
   - Significant dispersion (standard deviation). Busier airports and longer look-ahead times tended to have the largest standard deviations. For airports, standard deviations ranged from less than 1 flight (short look-ahead time at MCI) to 6 flights (longer look-ahead time at ORD and ATL). For sectors, standard deviations ranged from approximately 2 flights to 6 flights.

2. Linear regression algorithms considered in the study improved the accuracy of traffic demand predictions in comparison with current ETMS deterministic predictions. The major improvement (Model 2) was achieved by including the two predicted demands for adjacent 15-minute intervals (one for the preceding interval and another for the following one), along with the predicted demand for the 15-minute interval of interest that is currently used by ETMS.

3. Using additional variables in other regression models, such as active and proposed components in predicted demands, predicted demands for the 15-minute intervals 30-minute earlier and later of the interval of interest, and look-ahead time did not make significant contribution although they slightly improved the accuracy of demand

predictions achieved in model 2. Therefore, preference is given to model 2 due to its simplicity and efficiency.

4. For arrival airports, the use of predictions for adjacent intervals in the regression model (model 2) showed a substantial improvement (lower root mean square error (RMSE)) over current ETMS predictions. For the model with short LAT (15 min – 1 hr), RMSE dropped from 3.45 to 2.83. For the model with medium LAT (1 – 2 hr), RMSE dropped from 3.75 to 3.12.

5. A possible interpretation of the regression results is as follows:
   - If the output demand from the regression is higher than the deterministic prediction for the same interval, the regression may be capturing (a) migration of flights from another time interval to the interval of interest, and (b) pop-up flights.
   - If the output demand from the regression is lower than the deterministic prediction for the same interval, the regression may be reflecting the effect of (a) migration of flights to another time interval from the interval of interest, and (b) cancelled flights.

6. For sectors, model 2 also showed improved accuracy of demand predictions over current ETMS predictions, but not as significant as that for airports. For the model with short LAT (15 min – 1 hr), RMSE dropped from 3.75 to 3.43. For the model with medium LAT (1 – 2 hr), RMSE dropped from 4.32 to 3.86.

   Note that in ETMS, traffic demands for airports and sectors are defined differently. Airport aggregate demand for a 15-minute interval is measured by number of flights accommodated within the interval. For sectors, a 15-minute demand is defined as the maximum among fifteen one-minute flight counts within a 15-minute interval. It is worth mentioning that this metric is not a stable one because the maximum (as well as minimum) value of a set of random values is the least stable (robust) statistic among all order statistics and, hence, is harder to predict. Therefore, regression models for sector demand prediction based on maximum one-minute counts are less effective as the ones for airports. Further research is needed to improve the metric for traffic demand in sectors.

7. Model 2 was tested on three days of data that were not part of the original calibration set. It continued to show improved prediction accuracy.

8. The impact of model 2 on Monitor/Alert was examined. It was measured in terms of both number of alerts and stability (volatility) in identifying alerts. Airports and sectors become alerted when any of the 15-minute demand forecast during the planning horizon exceeds the corresponding alert threshold. Analysis was preformed on a two-hour planning horizon. Model 2 showed fewer alerts in comparison with the current ETMS functionality. The model improved stability (volatility) in identifying alerts in both airports and sectors. Volatility was measured by the number of times the predicted demand counts crossed the capacity threshold.

9. Results of this study should be considered as preliminary ones and further analysis would be needed on much larger data set. Nevertheless, the new regression models generally showed an improvement in demand predictions under uncertainty, but also demonstrated that substantial room for further improvement still remains, especially with respect to Monitor/Alert.

Section 6. Next Steps

This study focused on analysis of uncertainty in ETMS aggregate demand predictions as well as on improvement of prediction accuracy. The study was conducted on limited set of historical data and mainly during days without significant congestion. A next step would be to use proposed regression models for improving Monitor/Alert functions and estimate potential benefits for the days with predictions of significant congestion.

There are also two important areas of research that would be a direct continuation of what has been started in this study.

1. Further analysis of the error curves shown in Figure 1 and how they could be applied to traffic flow management. For example, how stable are these error curves over time? How do the error curves compare for different sectors or airports? Would it be useful if traffic managers could specify a probability threshold for alerts, e.g., show an alert if the probability of demand exceeding capacity is 30 percent or greater?
2. Analysis and characterization of uncertainty in predicted ETAs of individual flights for both airports and sectors. Use the results for probabilistic forecast of traffic demand with applications to Monitor/Alert.
3. Research on new metrics for sector demand that differ from the current ETMS metric (maximum of one-minute demand counts within a 15-minute interval). A candidate metric should be more robust and better predictable than current ETMS metric.

Section 7. References

[1] Wanke, Craig, Lixia Song, Stephen Zobell, Daniel Greenbaum and Sandeep Mulgund, "Probabilistic Congestion Management," 6th USA/Europe ATM R&D Seminar, Baltimore, 2005

www.ingramcontent.com/pod-product-compliance
Lightning Source LLC
Chambersburg PA
CBHW081803170526
45167CB00008B/3303